YOUR KNOWLEDGE HAS VALUE

- We will publish your bachelor's and master's thesis, essays and papers

- Your own eBook and book - sold worldwide in all relevant shops

- Earn money with each sale

Upload your text at www.GRIN.com
and publish for free

Bibliographic information published by the German National Library:

The German National Library lists this publication in the National Bibliography; detailed bibliographic data are available on the Internet at http://dnb.dnb.de .

This book is copyright material and must not be copied, reproduced, transferred, distributed, leased, licensed or publicly performed or used in any way except as specifically permitted in writing by the publishers, as allowed under the terms and conditions under which it was purchased or as strictly permitted by applicable copyright law. Any unauthorized distribution or use of this text may be a direct infringement of the author s and publisher s rights and those responsible may be liable in law accordingly.

Imprint:

Copyright © 2017 GRIN Verlag
Print and binding: Books on Demand GmbH, Norderstedt Germany
ISBN: 9783668630079

This book at GRIN:

https://www.grin.com/document/387313

Anchal Agarwal

The Glide-Shuffle Controversy in Silicon

GRIN Verlag

GRIN - Your knowledge has value

Since its foundation in 1998, GRIN has specialized in publishing academic texts by students, college teachers and other academics as e-book and printed book. The website www.grin.com is an ideal platform for presenting term papers, final papers, scientific essays, dissertations and specialist books.

Visit us on the internet:

http://www.grin.com/

http://www.facebook.com/grincom

http://www.twitter.com/grin_com

MATRL 288I, Final Class Project
University of California, Santa Barbara
Fall 2017

The Glide-Shuffle Controversy in Silicon

Anchal Agarwal

Abstract - Dislocations in Silicon have been a subject to intense studies in the last several decades. It is not only an interesting subject by itself, but is also important for understanding generic dislocation behaviors in a wider class of materials. In the 1970s, researchers had concluded that glide sets can move more easily than shuffle sets via experiments and theoretical calculations. It became widely accepted then that plastic deformation occurs by the motion of partial dislocations in the glide planes of diamond or zinc blende structures. TEM images confirmed this solidarity by showing the motion of dislocations under stress. In 1998, some researchers working on InP found that at very low temperatures (77 K) and high hydrostatic pressure, non-dissociated dislocations move in shuffle planes. It was also subsequently shown that a shuffle-set dislocation has a lower Peierls stress than glide-set partial dislocation. Other calculations debunked older models such as the Peierls-Nabarro model and showed that shuffle set-dislocations always move faster than glide sets. It has since broiled into a highly debated issue with a number of papers supporting either side. This paper attempts to give an overview of most of the seminal papers written on this topic and some newer work.

It was earlier assumed that dislocations which cause plastic flow in semiconductor crystals with diamond type structures were of the shuffle type, even though theoretically, both 'glide' set and the 'shuffle' set of dislocations (see figure 1) can dissociate in the diamond cubic structure [1]. The main reason for this assumption was that moving a glide set dislocation requires three times as many atomic bonds to be cut as moving a shuffle set dislocation [2].

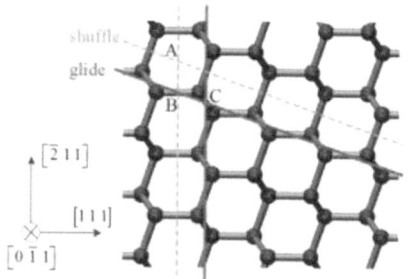

Figure 1: Glide and shuffle planes in Silicon

Early observations on dislocation networks contradicted this assumption and showed that dislocations were extended (glide type) [3]. This was challenged by Brooker et al., (1965) [4], who showed that unextended dislocations (shuffle type) and nodes could appear extended under particular diffracting conditions. They reported that observations on a wide variety of silicon specimens yielded no evidence for extended nodes. In an attempt to resolve this question, dislocations in silicon were examined using a weak-beam technique by

Cockayne et al. (1969) [5] (see figure 2). This provides a method of imaging dislocations as narrow peaks, typically 15 to 20 Å across, with an intensity many times that of the background. It greatly increased the resolution of detail compared to conventional dark-field images with the crystal close to the Bragg reflecting condition, where image widths were of the order of 100 Å. The weak-beam method was first used to resolve partial dislocations in copper-aluminium alloys and to measure their separation accurately. Later, it was applied to Silicon [6]. The images showed that dislocations at rest are extended, i.e., of the glide set. They also concluded that glide dislocations belong to the aB or that nucleation occurs easily even at room temperature [7]. A number of other researchers showed similar experimental results [8 - 13] in diamond and zinc-blende structures, including in III-Vs. Dynamic observations using strong beams showed that moving dislocations in Silicon were also extended [14].

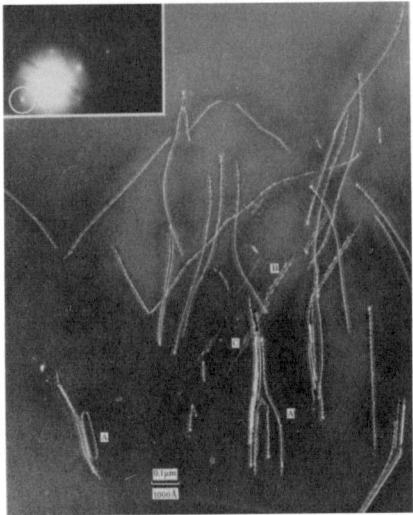

Figure 2: A weak-beam 220 dark-field image of the same area as in (a), showing a considerable increase in the resolution of dislocation detail [7].

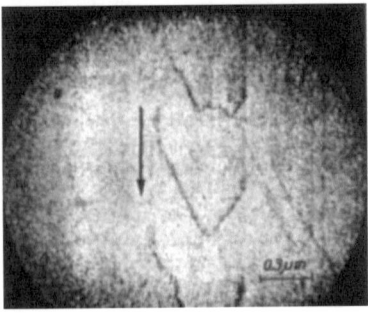

Figure 3: Dislocations moving at a low velocity at 700 °C [14].

Contrarily, lattice images of Ge and Si cores observed by an imaging technique using two tilted beams show that the dislocation core to be quite narrow, almost unextended [15 – 19]. Philips et al. [19] used a (Ill) lattice fringe image (see figure 4) of a 60° dislocation in a (110) foil from twisted germanium crystal to extract the dislocation width from the Peierls-Nabarro description. This extracted width had a value of 5.2 Å, much lower than dissociation width of 60 Å as determined by Ray and Cockayne. They reasoned that a dislocation dissociation of 60 Å should be readily visible their lattice fringe image, if present. They proposed three possibilities:

- A complete dislocation in the shuffle set is associated with a ribbon of stacking fault bounded by two Shockley partials of opposite sign in the glide set. One of these partials should then be near the core of the complete dislocation, the other being 60 Å away. They ruled out this model because of the complete symmetry of the observed displacement field about the terminating fringe.

- Deformation at elevated temperature produced two sets of dislocations, dissociated ones in the glide set and undissociated ones in the shuffle set. The dislocation imaged in Fig. 4 would then belong the shuffle set, the ones measured by Ray and Cockayne to the glide set.

- A further possibility is that the dissociated dislocation segments measured by Ray and Cockayne were not in the equilibrium state because of strong internal stresses acting on them.

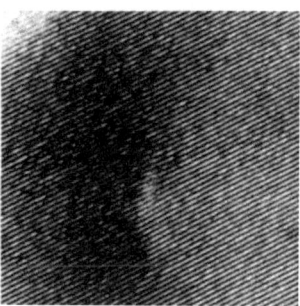

Figure 4: (Ill) lattice plane image of a 60° dislocation in a (110) foil from twisted germanium crystal. Fringe spacing is 3.266 Å [19].

Development of Axial beam imaging techniques in the 1970s made it possible to take more direct and exact lattice images. Sato et al. [20] used high voltage electron microscopy (accelerating voltage of 1000 kV) to image extended dislocations of 60° and screw types in deformed silicon crystals. All the dislocations observed by them were extended (shown in figure 5). In-situ transmission electron microscopy observations [21] demonstrated unequivocally that dislocations in silicon are dissociated into Shockley partial pairs which move without the need for diffusion, indicating that dislocation motion occurs exclusively on glide planes.

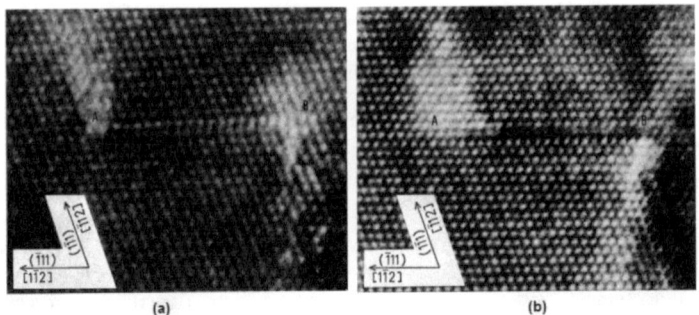

Figure 5: (a) An extended 60° dislocation and (b) An extended screw dislocation in Silicon imaged by high voltage electron microscopy [20].

On the other hand, more detailed theoretical work in the 90s confirmed Shockley's hypothesis (1953) – that breaking of gliding plane bonds is a dramatically higher energy process. Calculations of the Peierls stress for shuffle and glide dislocations using the Peierls-Nabarro (PN) model by Duesbery et al. [22, 23] with generalized stacking-fault energies indicating values nearly an order of magnitude higher for glide partial dislocations than for shuffle dislocations. Samuels and Roberts [24] made detailed measurements of the temperature dependence of the Peierls stress in silicon. The Peierls stress at the ductile-brittle transition temperature of 820 K is 0.0073 µ. Linear extrapolation of their results to 0 K, the appropriate limit for rigid-dislocation calculations, gives a value of 0.15 µ, roughly midway between the atomistic shuffle and glide values cited by Duesbery. This lend some experimental credit to the theoretical calculations, but furthered the glide-shuffle controversy. The observed free motion of Shockley partials led to the hypothesis that it is the dislocation dissociation itself which is responsible for the discrepancy.

Duesbery [25] proposed that in materials which exhibit significant dislocation-lattice coupling, the dislocations might not move by rigid translation (except near 0 K), but rather by the nucleation and propagation of kink pairs (mechanism shown in figure 6). A dislocation lying along a low-energy direction develops a thermally activated bulge (figure 6a) which under suitable conditions may reach the adjacent low-energy site (figure 6b), at which point the high-energy arms of the bulge, known as kinks, can move apart (figure 6c), dragging the entire dislocation through a unit translation.

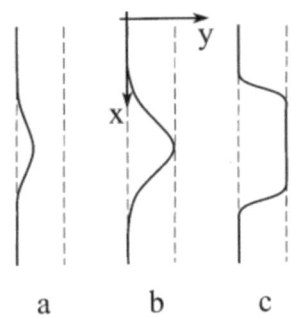

Figure 6: The kink pair mechanism for the motion of dislocations, as proposed by Duesbery [25].

Using an expression for the kink pair activation energy [26]:

$$U_{kp} = \int_{y_0}^{y_m} \left\{ W(y)^2 - \left[W(y_0) + \tau b(y - y_0) \right]^2 \right\}^{1/2} dy,$$

, Duesbery found that in usual range of stress ($\tau/\mu < 0.01$), the activation energy of partial glide kinks is favored over those of perfect shuffle kinks (as shown in figure 7). He concluded that for low stress, partial dislocations located in glide planes should be the more mobile species.

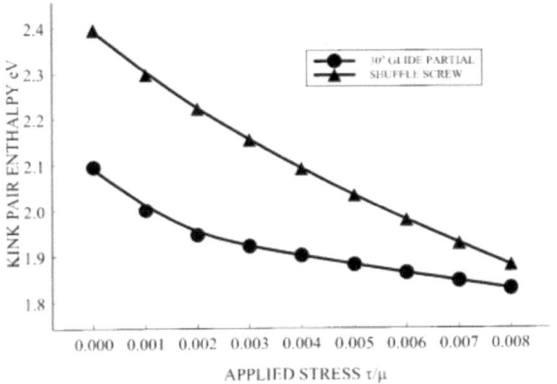

Figure 7: Kink pair activation energies in Silicon for a 30° glide partial (circles) and screw shuffle (triangle) dislocation as calculated by Duesbery and Joos [25].

The kink mechanism of glide-set dislocation was extensively investigated by Bulatov et al. [27, 28, 29] using atomistic simulation. They showed that complicated bond reconstructions on the dislocation core and kink play an important role in dislocation mobility. Cai estimated a kink-formation energy at $E_k = 0.728$ eV and a kink migration energy at $W_m = 0.022$ eV using SW potential [30]. The resulting effective activation energy for double-kink nucleation defined by $Q_{nucl} = 2E_k + W_m$ (~ 1.5 eV) was much smaller than that of 30 degree partial (2.6 eV) on glide set plane and of the Peierls-Nabarro model (2.4 eV) estimated by Dusbery. This indicated that shuffle-set dislocation always move faster than glide-set dislocations under all temperature and stress conditions. Results are summarized in Table 1.

Dislocation		σ_b (μ)	σ_p (μ)
glide	60° screw	dissociation	15.63 19.66
shuffle	60° screw	0.080 0.091	0.076 0.103
glide	90° 30°	0.27 0.33	0.450 0.561

Table 1: Comparison of Peierls stresses in Silicon obtained from the Peierls-Nabarro model [30].

A more recent work by Pizzagalli et al. [31] confirmed that shuffle sets should be more mobile over the entire stress range, not just low stress and low temperature, also contradicting Duesbery and Joos. Pizzagalli considered the velocity of a dislocation as being proportional to $e^{-Q/kT}$, Q being the activation energy of the process, and used a result by Hirth and Lothe [32] to calculate Q. They said that when the characteristic length of the dislocation segment is smaller than the average distance between thermal kinks, $Q = F^* + W_m$ (regime R_1). Otherwise, Q is equal to $F^*/2 + W_m$ (regime R_2), where F^* is given by:

$$F^* = 2\left[F_k - \sqrt{Kb^3h^3}\, \sigma^{1/2}\right]$$

Figure 8 shows the variance of Q as a function of applied stress σ (μ) for R_1 and R_2 regimes. The big difference from the original data by Duesbery and Joos is the absence of an intersection between the 30° partial and screw dislocations curves. This suggests that the shuffle screw dislocation should be more mobile than the 30° glide partial dislocation for all stresses.

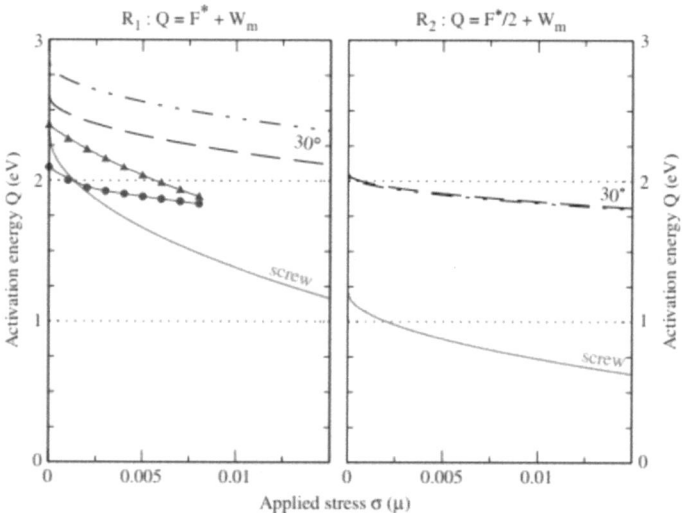

Figure 8: Activation energy for the thermally activated motion of dislocation as a function of applied stress, determined by Pizzagalli et al. [31].

Saka et al. [33] observed that the shuffle set of dislocations were transformed into a glide set of dislocations at around 400°C. The reasoning for this was that 400°C, the dislocations were dissociated into two partials (seen by TEM), whereas at room temperature, no evidence for the dissociation was obtained. Note that there is no direct method to distinguish between shuffle and glide sets via electron microscopy. At higher temperatures (> 750°C), screw dislocations in the glide set were transformed into helices. The TEM images taken during the heating process is shown in figure 9.

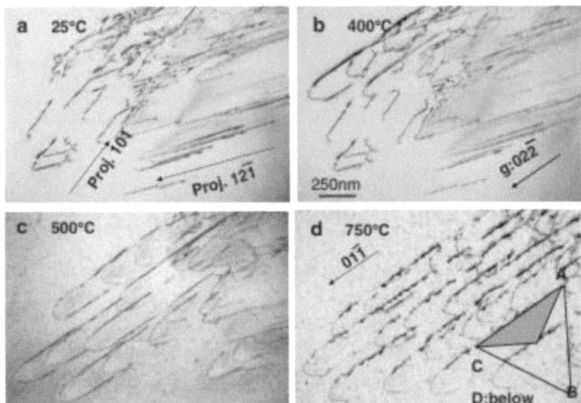

Figure 9: Change in the configuration of the initially shuffle set of dislocations during heating, seen via TEM by Saka et al. [33].

My understanding of the entire controversy is that a crystal deformed at low temperatures and large stresses deforms primarily by the motion of shuffle dislocations, while at high temperatures and lower stress, dislocations nucleating and moving in the glide plane cause most of the plasticity. This is because at low temperatures dislocations are generally observed to be un-dissociated, while at high temperatures the situation is largely reversed, indicating transition into gliding sets. The transformation from shuffle to glide is credited to interaction with the supersaturation of interstitials, but a number of counter-arguments weaken that claim. Understanding the mechanism of the shuffle to glide transformation, if true, requires further work.

References

[1] Hirth, J. P., and J. Lothe. "Theory of Dislocations, 780 pp." (1968).

[2] Shockley, W. (1953, January). Dislocations and edge states in the diamond crystal structure. In Physical Review (Vol. 91, No. 1, pp. 228-228).

[3] Aerts, E., Delavignette, P., Siems, R., & Amelinckx, S. (1962). Stacking fault energy in silicon. Journal of Applied Physics, 33(10), 3078-3080.

[4] Booker, G. R., & Brown, L. M. (1965). Observations on dislocation nodes in silicon. Philosophical Magazine, 11(114), 1315-1319.

[5] D. J. H. Cockayne, I. L. F. Ray & M. J. Whelan (1969) Investigations of dislocation strain fields using weak beams, The Philosophical Magazine: A Journal of Theoretical Experimental and Applied Physics, 20:168, 1265-1270

[6] I. L. F. Ray & D. J. H. Cockayne (1970). The observation of dissociated dislocations in silicon, The Philosophical Magazine: A Journal of Theoretical Experimental and Applied Physics, 22:178, 853-856

[7] Ray, I. L. F., & Cockayne, D. J. H. (1971, December). The dissociation of dislocations in silicon. In Proceedings of the Royal Society of London A: Mathematical, Physical and Engineering Sciences (Vol. 325, No. 1563, pp. 543-554). The Royal Society.

[8] Häussermann, F., & Schaumburg, H. (1973). Extended dislocations in germanium. Philosophical Magazine, 27(3), 745-751.

[9] Gomez, A., Cockayne, D. J. H., Hirsch, P. B., & Vitek, V. (1975). Dissociation of near-screw dislocations in germanium and silicon. Philosophical Magazine, 31(1), 105-113.

[10] Packeiser, G., & Haasen, P. (1977). Constrictions in the stacking faults of dislocations in germanium. Philosophical Magazine, 35(3), 821-827.

[11] Gottschalk, H., Patzer, G., & Alexander, H. (1978). Stacking fault energy and ionicity of cubic III–V compounds. physica status solidi (a), 45(1), 207-217.

[12] K. Wessel & H. Alexander (1977) On the mobility of partial dislocations in silicon, The Philosophical Magazine: A Journal of Theoretical Experimental and Applied Physics, 35:6, 1523-1536

[13] Gomez, A. M., & Hirsch, P. B. (1977). On the mobility of dislocations in germanium and silicon. Philosophical Magazine, 36(1), 169-179.

[14] Sumino, K., & Sato, M. (1979). In-situ HVEM Observations of Dislocation Processes during High Temperature Deformation of Silicon Crystals. Crystal Research and Technology, 14(11), 1343-1350.

[15] Parsons, J. R., & Hoelke, C. W. (1969). Crystal-Lattice Images of End-On Dislocations in Deformed Aluminum. Journal of Applied Physics, 40(2), 866-872.

[16] Parsons, J. R., Rainville, M., & Hoelke, C. W. (1970). Influence of crystalline defects on 2-beam crystal lattice images-Experimental. Philosophical Magazine, 21(174), 1105-1117.

[17] Phillips, V. A., & Hugo, J. A. (1970). The resolution of lattice planes and lattice defects in semiconductors including observations on boundaries. Acta Metallurgica, 18(1), 123-135.

[18] Phillips, V. A. (1972). Lattice resolution observations on the structure of twin boundaries, faults and dislocations in epitaxial silicon. Acta Metallurgica, 20(10), 1143-1156.

[19] Phillips, V. A., & Wagner, R. (1973). Structure of dislocations in germanium. Journal of Applied Physics, 44(9), 4252-4254.

[20] Sato, M., Hiraga, K., & Sumino, K. (1980). HVEM structure images of extended 60-and screw dislocations in silicon. Japanese Journal of Applied Physics, 19(3), L155.

[21] Alexander, H. (1986). Dislocations in Solids. Ed. FRN Nabarro, 7.

[22] Joós, B., Ren, Q., & Duesbery, M. S. (1994). Peierls-Nabarro model of dislocations in silicon with generalized stacking-fault restoring forces. Physical Review B, 50(9), 5890.

[23] Kaxiras, E., & Duesbery, M. S. (1993). Free energies of generalized stacking faults in Si and implications for the brittle-ductile transition. Physical review letters, 70(24), 3752.

[24] Samuels, J., & Roberts, S. G. (1989, January). The brittle-ductile transition in silicon. I. Experiments. In Proceedings of the royal society of London A: mathematical, physical and engineering sciences (Vol. 421, No. 1860, pp. 1-23). The Royal Society.

[25] Duesbery, M. S. (1996). Dislocation motion in silicon: the shuffle-glide controversy. Philosophical magazine letters, 74(4), 253-2

[26] Dorn, J. E., & Rajnak, S. (1964). Nucleation of kink pairs and the Peierls' mechanism of plastic deformation. Trans. Aime, 230(8), 1052-1064.

[27] Bulatov, V. V., Yip, S., & Argon, A. S. (1995). Atomic modes of dislocation mobility in silicon. Philosophical Magazine A, 72(2), 453-496.

[28] Bulatov, V. V., Justo, J. F., Cai, W., Yip, S., Argon, A. S., Lenosky, T., ... & Rubia, T. D. D. L. (2001). Parameter-free modelling of dislocation motion: the case of silicon. Philosophical Magazine A, 81(5), 1257-1281.

[29] Bulatov, V. V. (2001). Bottomless complexity of core structure and kink mechanisms of dislocation motion in silicon. Scripta materialia, 45(11), 1247-1252.

[30] W. Cai, Doctoral thesis, MIT, (2001).

[31] Pizzagalli, L., & Beauchamp, P. (2008). Dislocation motion in silicon: the shuffle-glide controversy revisited. Philosophical Magazine Letters, 88(6), 421-427.

[32] Hirth, J. P., & Lothe, J. (1982). Theory of dislocations.

[33] Saka, H., Yamamoto, K., Arai, S., & Kuroda, K. (2006). In-situ TEM observation of transformation of dislocations from shuffle to glide sets in Si under supersaturation of interstitials. Philosophical Magazine, 86(29-31), 4841-4850.

[34] Li, Z., & Picu, R. C. (2013). Shuffle-glide dislocation transformation in Si. Journal of Applied Physics, 113(8), 083.

YOUR KNOWLEDGE HAS VALUE

- We will publish your bachelor's and master's thesis, essays and papers

- Your own eBook and book - sold worldwide in all relevant shops

- Earn money with each sale

Upload your text at www.GRIN.com and publish for free